NIGHT VISION

Pippa Goldschmidt lives in Edinburgh and Berlin. She has a background in astronomy and is particularly interested in writing about science. Most recently she co-edited (with Drs Gill Haddow and Fadhila Mazanderani) *Uncanny Bodies*, a specially commissioned anthology of fiction and essays responding to Freud's uncanny. Her work has been broadcast on BBC Radio 4 and published in *ArtReview, Tamarind, BBC Sky At Night, Mslexia, Times Literary Supplement,* and *Magma*.

Also by Pippa Goldschmidt

The Need for Better Regulation of Outer Space	(Freight Books, 2015)
The Falling Sky	(Freight Books, 2013)

ISBN: 978-1-916938-03-8

Cover designed by Aaron Kent

Edited & Typeset by Aaron Kent

Broken Sleep Books Ltd
Rhydwen
Talgarreg
Ceredigion
SA44 4HB

Broken Sleep Books Ltd
Fair View
St Georges Road
Cornwall
PL26 7YH

Night Vision

Pippa Goldschmidt

Broken Sleep Books

The night skies of my London childhood were drowned in street-light and when I looked upwards, all I could ever see was the Moon with its features blurred by clouds and pollution. The planets and the stars were neither more nor less than pictures in books; I learnt the constellations from small black and white photos in one of Patrick Moore's guides to astronomy, the grain of the film still visible on the page. My children's encyclopædia had an entry for the Universe showing an expanding balloon covered with black dots for galaxies. I had no real idea what 'Universe' or 'galaxies' meant, but I was drawn to the words themselves. The encyclopædia also told me that the Sun was actually a star and furthermore it was a radio star, and I marvelled at this apparent connection with the metal box that bristled with buttons and dials, and sat on the kitchen counter, its antenna pointing up at the kitchen ceiling. What could the words 'radio' and 'star' possibly have in common? But I loved the phrase 'radio star', and would repeat it to myself, perhaps because it hinted at another sort of reality, a different type of truth encoded in the words. For me, 'radio star' was a poem.

When I left London at the age of eighteen and moved to Leeds to study physics and astronomy, I realised that cities did not stretch endlessly to the horizon and that the night sky was not tinted orange. Here, I was taught how to look properly at this sky, to train my eyes to become dark-adapted by turning away from the surrounding buildings and car headlamps, and patiently waiting before I could make out dim stars, the merest twinkles of light.

This mini apprenticeship I underwent each cloudless evening gave me a chance to appreciate the stellar landscape. I learnt to navigate around the sky from the ladle of the Plough to the Pole star. This new-found ability revealed for me not only well-known constellations such as Orion or Cassiopeia, but also the smaller, fainter targets of my academic enquiry. Ever since, I have associated the Hyades stellar cluster with standing on the roof of Leeds University's physics department, and getting a crick in my neck from staring at this V-shape group of stars located in the constellation of Taurus. After I learnt to find the Hyades I was allowed to look at them through the departmental telescope which separated out the small cluster into individual stars, like looking at a crowd of people and being able to identify their faces.

Using a telescope was not just about learning to see objects in the sky, it involved a whole collection of new sensory experiences such as attempting to turn the focus dial of the eyepiece with hands so cold I could barely feel the ridges on the metal. Not breathing anywhere near the lens because the resulting condensation took minutes to clear. Not getting distracted by the reflection of my own eye regarding me, the magnified eyelashes and iris superimposed on the tiny circle of sky. Swigging from flasks of strong tea and eating peanut butter sandwiches at three o'clock in the morning to try and keep up my flagging energy. Waiting for the sky to clear when the stars were blocked from view by the inevitable clouds.

The night sky soaked into my bone-cold hands and feet, and made my neck hurt from peering through the eye piece or tilting my face upwards. Night seeped into day, made my head ache with tiredness when I returned to my bedroom and tried to get to sleep. It took up residence throughout my body, a sort of ever-present darkness that balanced out the fluorescent-lit days I spent sitting in labs and lecture halls. If those days could be stressful because of the desires and the uncertainties of being a student, the night sky provided a ballast. During this time of measuring the Hyades I started a relationship with a boy, and when I had to say goodbye to him each dusk-filled afternoon in late autumn (the time of the year when the Hyades are at their zenith and easiest to observe) he would sing 'Starry, starry night' to me.

To view the constellations night after night is to experience a kaleidoscope of associations. In the northern constellation of Andromeda there is a faint thumb-print of light, one of a very few galaxies that can be seen by the naked eye without a telescope. In Greek mythology, the gods chain Andromeda to rocks as a punishment for the hubris of her mother Cassiopeia, before she is saved from a threatening sea monster by the heroic Perseus who then marries her. The mythological Andromeda is a passive woman, seemingly without any agency over her own life and fate, and the galaxy Andromeda is hurtling towards the Milky Way, nothing can stop the two from colliding in a few billion years. This flickering to and fro between myth and science is a reminder that the sky is not a passive backdrop to our human activities, it is a theatre and a generator of thought.

As I progressed through my degree, I learned that the Sun emitted not just visible light but also radio waves which peaked and troughed in cycles I could plot on a piece of paper and explain with a mathematical equation. 'Radio star' became a graph of curved lines and data points as well as a poem.

Standing on the roof of the physics department and preparing to look at the Hyades, I would sometimes spin round and round and watch the stars rotate the opposite way. The Universe was a giant spinning top and I was at its centre. I learned that this was not true, the Universe had no centre. The boy I had fallen in love with told me he thought he might be attracted to other boys.

Or not. He wasn't sure, but he would still sing to me while he tried to decide. The Universe was expanding, everything moving further away from everything else. The gaps between the galaxies would only grow larger as time flowed from the past through our present, and on into the future.

These gaps that we see between stars and galaxies are not easily explained. In principle, an infinitely large, infinitely old Universe should be as least as bright in every single direction as the surface of the Sun and wherever you look at night-time you should see light from one of the infinite number of stars. But this is not the case because the sky appears to be dark at night; individual stars are surrounded by apparently empty space. The boy left, then he came back to me. Why were we surrounded by so much darkness? Such an apparently obvious question is problematic to answer. For my degree I wrote an essay about the many possible explanations of this paradox of the dark sky. I wrote about the genius required to identify what is uncanny about the most obvious observation anyone can make of the night. Based on other people's more coherently expressed arguments, I wrote that this only appears to be a paradox because of our apparently reasonable assumptions, and in fact the Universe cannot be both infinitely large and old. Something has to give. According to the Big Bang theory, the Universe began in an explosion 13.7 billion years ago and has been expanding ever since; in this scenario there has not been enough time for the light from the most distant stars to reach us on Earth and that is one of the reasons for the darkness of the night sky. We will never be able to see all there is to see before it is carried away from us.

After I left Leeds I moved further north to Edinburgh to do a PhD in astronomy, where I spent four years at the grandly-titled 'Royal Observatory Edinburgh'. But people couldn't do any actual observing from here because the telescopes had all been decommissioned many years beforehand, due to the combination of Scottish weather and light pollution from the city. The telescopes remained in their domes at the top of the observatory, vast constructions of metal struts and silvered mirrors; memorials to long forgotten observations and disproved theories. The domes were always hushed and dark, like churches where people didn't bother going to pray anymore. Now the astronomers sat in offices and summoned up their data on computer screens.

My research project concerned quasars; rare and luminous centres of very distant galaxies, and thought to be powered by super-massive black holes. To explore these quasars in a properly scientific and thoroughly systematic fashion, I built up a catalogue of a section of the sky using photographs. We were in an era of rapidly changing technology, and digital cameras were starting to be developed, but it was still quicker to use the nineteenth century technology of glass photographic plates to make large surveys of the sky. These plates could be viewed on special light tables, and we were taught to hold each square of glass by its edges to avoid damaging the delicate emulsion before laying it flat on the glowing surface of the table. I spent a lot of time peering at plates, looking down over a scatter of stars and galaxies. The photographs were actually negatives in which the darkness of the night sky was reversed and appeared white, freckled black by stars in our own Milky Way Galaxy and interspersed with other, more distant galaxies as seemingly delicate as puffs of smoke. If you looked really closely you could see the individual grains of the emulsion, the fabric of the image. But I wasn't able to spend much time peering at the photographic plates, because each one contained images of upwards of 100,000 astronomical objects and so it was more efficient to scan them and create databases of the information. Most of the time the plates themselves remained stacked, unseen, in the observatory plate library whose floor had to be specially reinforced to cope with the combined weight of all that glass.

This was an era of transition not just in technology but also in the science itself. The Universe was taking a more definite shape, its characteristics were being ever better measured and defined. At the start of my PhD we didn't know the speed of the Universe's expansion, expressed as the value of the Hubble constant, to any better than 50% accuracy but upper and lower boundaries were swiftly becoming narrower. During my PhD the boy and I got engaged to be married, it seemed like a grown-up thing to do. I contributed to those lessening of possibilities and increases in certainties for the Universe. I gradually became better at statistics, and learnt how to wrote computer code that analysed data. And as we transitioned from photographic plates to digital images, analogue got left behind, and the stars and galaxies were now pixelated into little squares with hard and definite edges.

As I coded and calculated, I learnt about the Cosmological Principle; 'we are not special'. This is an assumption that Earth-bound observations are not unique to us, and what ours tell us about the Universe is consistent with possible ones taken from elsewhere. The Principle is only true on large scales, but it is this assumption that allows us to do astronomy, and to say anything of interest about the Universe as a whole.

The Cosmological Principle is a more general reworking of the Copernican Principle, stemming from the demotion of the Earth from centre of the known world to one planet amongst many that orbit the Sun. Since the publication of Copernicus' heliocentric model in 1543, astronomers have been successively relegating the Earth and its star; now we know that the Sun is just an average-sized middle-aged star, one of billions within the Milky Way. And since Edwin Hubble's work in the 1920s measuring distances to nearby galaxies, we have known that the Milky Way itself is one of many, many billions of galaxies throughout the Universe. Space missions show that the Earth is a smallish rocky planet, similar in many ways to Mercury, Venus and Mars. In the past twenty years thousands of 'exoplanets' orbiting other stars have been detected. Some forms of life must exist on those planets, maybe even on objects elsewhere within our solar system. Our uniqueness is illusory. We are not special.

When I was at the observatory I also learnt to stop using the word 'I' when writing about my work. The ideal academic paper is not about its author who should be invisible to the readers and who writes sentences like 'The stars were observed'. 'I' did not observe them because I am not special, I am the archetypal and generic observer, I could be anywhere in the Universe. It was perhaps an accident that I happened to be in an office at the top of a tower built in the latter part of the nineteenth century, and positioned right on the edge of the city - this edge sometimes reminding me of the boundary between light and dark that is visible during an eclipse. Look south from the tower and you can see fields, open expanses of green, and distant hills. Look north and you can see the grey rooftops, the castle and Arthur's Seat. On a sunny day, the distant Firth of Forth glitters in the north-east. The observatory itself, with its towers, telescopes, library of plate glass photographs and hoard of ancient astronomical texts, seemed like a pretty special and unique place to me, but perhaps that was also an illusion.

Now I wore an engagement ring on my finger, the silver created in unknown stars when they exploded at the end of their lives. Because of his work, the boy went away for months at a time, he said it was the only way to get ahead. The Moon would swing back and forth across the sky and I learned to be by myself.

Looking at stars without a telescope requires you to stand outside in the dark. You must wait for your eyes to acclimatise, to become more sensitive to the faintest light. It takes about fifteen minutes on average for this to happen, fifteen minutes of letting your gaze settle on patches of blank sky, waiting for them to reveal their stars. When your eyes eventually achieve this 'dark-adaptation' it feels like a small victory against the city. And once achieved, dark-adaptation must be protected, you must avoid glancing in the direction of any artificial lights, particularly those – like phone screens – that cast a blue light. Even so, I remained grateful for streetlights that allowed me to walk to and from the Observatory in the dark days of Edinburgh's winters.

But you will never see a quasar with your naked eye. Although I knew exactly where to look in the sky, the brightest quasar is about a hundred times too faint to be seen. A quasar is always mediated by photograph, by metal, by glass, by silver emulsion, by computer screen and by pixel.

During my PhD the Hubble Space Telescope was launched. When it started to produce images, they were unlike any we had seen before. We felt ourselves to be looking directly through its eyepiece far above the Earth. The clarity! The beauty! More prosaically, you couldn't see the outlines of the pixels themselves; the digital technology was finally, as we had always been promised, invisible. This was what astronomy was meant to be. Just us and the swirls of coloured gas around young stars, distant galaxies glittering like jewels. The error bars of the Hubble constant lessened until they were almost zero. Now we knew exactly how fast the Universe was expanding and consequently how fast we were travelling away from every other galaxy. This was what the Hubble had been built for, really, to give us one number. A number we could be certain of.

Around this time the boy and I got married and then, a year or so afterwards, we split up. We had learnt very little about each other while we were together, apart from the fact that we were better off apart. A paradox. Perhaps some paradoxes have to be experienced rather than studied.

The hallmark of a scientific observation is that it is capable of being reproduced in other places and by other observers; nothing that 'I' observe should be unique to me. The Cosmological Principle haunted my work; I spent years as a PhD student and as a junior academic trying to figure out exactly what was special about my work. If it was supposed to be reproducible, then what was distinctive about it, what could I point to and say 'this is mine'? What, apart from the mistakes?

Because the telescopes in Edinburgh were no longer in use, whenever we needed to gather observations of the objects we were studying, we travelled abroad to observatories on the tops of mountains above the clouds in the Canary Islands, Australia and Chile. In the high altitude Atacama desert of northern Chile, where the air is thin and so dry that not even bacteria can survive, at night the mountainous landscape becomes insignificant as the sky above fills with brilliant stars. Here, the Milky Way runs across the sky like an ancient footpath through a field, and its centre is so bright and clear it casts a shadow behind me as I gaze up.

The link between the brilliance of the night sky in Chile and the lack of local human activity is obvious. No motorways pass through these mountains and the nearest village is hundreds of miles away. Astronomy is a passive science in which people seek to receive and detect tiny fragments of light, so telescopes and ob-servatories are designed to cause almost no disturbance to their immediate physical environments.

The European Southern Observatory (ESO) in Chile was first founded in the 1960s. Now there are around fifteen telescopes here, as well as support buildings, a residential place for people to sleep, a library, offices and a canteen. At any one time there might be a few hundred astronomers visiting the place, staying for a few days or several months. The land that the observatory is built on was purchased from the Chilean Government and is legally German.

ESO has been continually operational since 1969. On the night of the coup on 11 September 1973 instigated by General Pinochet against the democratically elected Marxist government headed by Salvador Allende, the subsequent imposition of mar-tial law and throughout the entire span of the repressive military regime, the telescopes kept collecting light, and the astronomers from various European countries kept arriving, recording images of distant stars and galaxies, and departing.

I first went to Chile in the spring of 1990, the year that Pinochet's regime officially came to an end (after the national plebiscite in 1988) and Chilean society started easing itself back into a democracy. Before I left Edinburgh, I was warned not to discuss politics with any Chileans, such as the people who worked at the observatory in a support capacity, because to do so might put them at risk. I use the passive voice "I was warned" because I cannot now remember who warned me. Was it even a specific voiced warning, or just a general silent understanding that one did not, should not, talk about these things? Although Pinochet had stepped down and the new government had peacefully assumed power, a lot of people still supported him and it was a very divided country. But I did as I was told and I kept my mouth shut. When I worked at the telescope and I gave the Chilean technician the co-ordinates for the objects I was interested in, we didn't talk about anything apart from food.

The telescopes had to be pointed away from the Sun and the Moon, otherwise the delicate equipment was at risk of being fried by the intensity of the light emitted from those relatively nearby objects. But the thing itself, the quasar, was an object in a different timeframe. I spent my life looking at and thinking about objects that had existed billions of years ago. They didn't exist 'now', all I could see of them was the remnants of light they once emitted. In this 'now' the technician and I sat side by side and gazed at the computer screens in silence.

The astronomers never talked about any of the politics or the obvious divisions in Chile, not even when we were back home in Edinburgh. We never commented on the ostentatious wealth in the suburbs of Santiago, where the observatory's administrative headquarters were located, nor on the nearby shanty towns built from sheets of corrugated iron. Neither did we comment on the fact that the astronomers were nearly all European and the support staff were Chilean. Or that the *residencia* in Santiago where we stayed while we waited to travel further north to the Observatory or fly back home, was a place that had the semblance of a colonial-era hotel in which we were served food by Chilean women dressed in old-fashioned black dresses and white aprons. This *residencia* was in the same neighbourhood as the elite military academy where Pinochet's coup had first been plotted; another silenced fact.

We were too intent on peering into the remotest aspects of the Universe to comment on what was going on around us. Perhaps the habit of collecting light without any means of influencing the sources of that light has rendered astronomers rather passive. ESO couldn't join in the celebrations around Chile's overthrow of its military dictatorship, because it had never actually condemned Pinochet or any of his murderous crimes. And it is not the only such observatory, there are several other American and Canadian funded and operated telescopes in the Andes, their domes gleaming in the desert like metallic mushrooms and looking so out of place in that environment they might be aliens.

Astronomy can seem esoteric, needing expensive equipment and yielding complex theories; perhaps it's an apparent luxury in an age of need. Actually it's essential for understanding how we relate to our surroundings both near and far. In the late 16th century Tycho Brahe spent many years observing Mars, constructing careful tables of that planet's position as it appeared to advance across the sky and then double back on itself for months at a time in a retrograde motion, before changing direction yet again. This seemingly complex dance was explained by Johannes Kepler in the early 17th century as being due to Mars (and by implication all the other planets) orbiting the Sun. Copernicus had already proposed this more than fifty years previously, but now Kepler showed that the geometrical form for these orbits that best fit the data was an ellipse and not – as had been assumed by Copernicus – a circle. Kepler was able to derive equations for planets' motions that accurately predicted the way they speed up as they approach the Sun and slow down as they recede from it.

Kepler's work quantifying astronomical orbits was an essential precursor to Isaac Newton's realisation in the late 17th century that the same force governs the planets' motion around the Sun, a (hypothetical) satellite orbiting the Earth and an apple falling from a tree to the ground. Newton's work conceptualising this universal force of gravity makes the link between our everyday world of apples and observations of light specks in the sky. We're drawn closer to outer space.

But Newton's outer space is a giant chessboard where both space and time are needed to explain the motions of astronomical objects – but are not themselves affected. Over two hundred years later, Newton's universal force of gravity was superseded by Einstein's general theory of relativity and its concept of space-time, a more interactive connecting material than Newton's universal space and absolute time. Space-time influences the movements of objects in it, but is altered itself by those objects. Einstein used this concept to explain a known oddity in the observed orbit of Mercury as it passes through space-time warped by the nearby Sun.

Although astronomers can only directly detect light, we know that darkness is crucial to the Universe. It is the unseen that holds together the seen. In the 1970s, Vera Rubin observed how fast spiral galaxies seemed to be spinning relative to their visible mass, and concluded that there must be some form of invisible 'dark' matter surrounding them that stops them from flying apart. But we still have little idea about what this dark matter actually is; we have never detected it directly, cupped it in our hands or weighed it in a test tube. It is always travelling through us on its way to somewhere else, never stopping.

Western constellations tend to be all about the light, but those created in other societies recognise the role of darkness. The

Australian aboriginal constellations are often defined by their lack of light; the Emu constellation is made up of opaque dust clouds in the central band of the Milky Way.

The fabric of space-time exists in a seemingly endlessly symbiotic relationship with planets, stars and galaxies. We move because of the influence of all the other objects in the Universe, or, as the Ubuntu saying has it, I am because you are. Now we are truly embedded in outer space. And we're affected by those outer space objects in real and measurable ways. On North Ronaldsay, the northernmost island in the Orkney archipelago, lives a flock of sheep owned collectively by the small community. In the past, and before a regular and reliable boat service was established, these people needed every square inch of the island to cultivate food. They couldn't afford to grow grass for the sheep, consequently in the nineteenth century the entire flock were banished to the coast beyond a perimeter wall which was built to enclose and protect the precious food crops.

Sheep are ruminant animals, chewing the cud at various intervals throughout the day. But the sheep on North Ronaldsay have adapted to eat seaweed, the only food available. They have to eat it when it becomes accessible to them at low tide, and conse-quently they have to chew the cud at high tide, twice a day, even as the timing of the tides slips around the earthly clock due to the Moon's monthly orbit. Their digestive systems have become at-tuned to this orbit and the subsequent shifting to and fro of the sea.

The sheep have themselves become a celestial clock.

Sheep in the sky are also a form of clock; we tell the time in astronomy by using Aries as the start of Spring, the vernal equinox, the moment each year at which the Sun crosses from the southern to the northern hemisphere. The first point of Aries is designated as a celestial prime meridian, the zero point for measuring the east-west position of objects in the sky.

Outer space is many layered. Kepler spent years trying to understand the movements of the planets before he managed to derive mathematical laws of their motion. But he earned much of his livelihood by casting horoscopes, linking the planets' apparent wanderings across the sky to the fates of individuals. He also became obsessed with making a connection between planetary orbits and the concept of Platonic solids, such as the cube and the tetrahedron, in which each face and angle are the same. He thought the distances between planets should follow the gaps between these solids if they are nested within each other. As with modern chemistry's murky past in the fume-filled dens of medieval alchemists struggling to turn base metal into gold, so modern astronomy has emerged from a decidedly non-scientific mixture of beliefs.

When I was a teenager my mother became interested in astrology, attending weekly workshops to learn how to draw 'natal charts' that showed the planets' positions in the sky at the time of a person's birth. It irritated me when I saw her studying the ephemeris, searching for some sort of secret truth, and apparently believing this arbitrary system of associating planets with people's destinies. Everyone knew that astrology was just a relic from the past, I told her, from an age when people didn't know any better. In her version of reality, the planets revolved the Earth which was still fixed in its pre-scientific location at the centre of the known Universe. She ignored my arguments, and enjoyed annoying me by telling me that my misgivings about astrology could be predicted in my horoscope.

Just as I was preparing to go to university, Halley's comet appeared after orbiting around the solar system for the previous 76 years. But after all the media hype it was a disappointing spectacle, nothing more than a faint smear in the sky and (of course) almost completely invisible from London. At that time I was much more interested in quantum physics, in the bizarre behaviour of light and subatomic particles that defied common sense and everyday experience. Perhaps I did not want to believe that my mother could predict my future in such a non-scientific way. But as it turns out, an astrologer gave birth to an astronomer.

Not long after I started my degree I became intrigued by the idea of a distant celestial object being simultaneously in the sky above me and also on the graph paper in my lab book. For one of my first assignments, I was required to calculate the amount of energy Jupiter receives from the Sun over a range of different frequencies, and compare this with what it emits. Jupiter is one of the brightest objects in the night sky, easily seen by the naked eye. But this is because, like all other planets, it only shines due to reflected light from the Sun, it can't generate any of its own.

However, with a simple calculation, I proved that Jupiter was brighter in certain wavelengths than it should be if it was solely relying on the Sun's light. It did indeed generate its own light, albeit low energy light in the form of radio waves. If the Sun is a radio star, Jupiter is a radio planet. And if it can generate its own energy, it's something akin to a failed star, eliding easy boundaries. I did not want to be easily understood either, not by my mother nor anyone else. I decided to study astronomy alongside physics.

Sometimes in astronomy we only see what we want to see. In the early twentieth century Percival Lowell made detailed observations of Mars, and claimed he'd found evidence that the surface was covered in 'canals'. Furthermore, these canals must be artificial and been built by an advanced civilisation trying to channel water from the planet's polar regions to its equatorial ones. Most other astronomers were sceptical but Lowell's claims were enormously influential on popular culture, and not completely disproved until the NASA and Soviet fly-by missions in the 1960s. Lowell also claimed to detect a large dark spot on the face of Venus, this was actually an image of his own pupil reflected into the telescope optics; for him outer space had become a mirror.

For me, it became an arena where I could convert images into databases of numbers and then into words. The quasars that I detected using the telescopes in Chile and elsewhere were alchemised by me from numbers into narratives, I wrote scientific papers speculating on how these objects first formed in the early Universe and what might have happened to them to make them much less common in more recent epochs. Together with other astronomers I talked about the birth and death of galaxies, quasars, black holes and stars, and about long-lived and short-lived objects, although I was never quite sure if this talk of birth, life and death was a metaphor, a shorthand for referring to the coalescing of gas into gravity-bound objects, and its subsequent dispersal. A metaphor that subconsciously helps us associate with the transient nature of those objects.

We have become used to seeing the sky depicted in beautiful images produced by the Hubble space telescope; images that revealed in more detail than ever before the shapes, colours and structures of galaxies, clusters and star-forming regions. But these images have been manipulated in a way that triggers an emotional response from us. The original detections made by the telescope are simply arrays of numbers recorded on a digital camera and converted to greyscale. These greyscale images are taken at different wavelengths which are then each assigned a colour, before being combined into a multi-colour image. There is a subjective choice in deciding which colour might correspond to each wavelength, but this choice is not apparent in the final, much publicised pictures.

The Stanford art historian Elizabeth Kessler has written about the way in which Hubble's images have been manipulated in a way that evokes the palette of nineteenth century paintings of the American West by artists such as Albert Bierstadt and Winslow Homer, this art attempted to convey the overwhelming scale and size of the new frontier to people who'd never laid eyes on it. This was art that prompted the sublime, that feeling of being overawed and overwhelmed when confronted by untamed Nature (and a Nature that is conspicuously empty of any indigenous peoples), and Hubble's images have the same effect. This in turn also reinforces the heroic aspect of our astronomy – it's us alone facing the mighty and awesome cosmos; the last and largest frontier, but we can tame it with our human-sized technology.

Unlike my mother's friendly group of astrologers, who gathered around our dining table to pore over charts and tables of numbers and who talked unselfconsciously about the colour of their auras, the 'heroic' space in which I worked was almost entirely gendered male. After my PhD I got a junior post in the astrophysics unit at Imperial college, where people phoned 'Dr Goldschmidt' and after I replied, would fall silent while they tried not to sound surprised at hearing a woman's voice. I had a grant supervisor who was supposed to take an interest in my work, but he never spoke to me. I had a manager who wanted me to do his work, and not mine. The female toilet was rumoured to be the old storage cupboard of a radioactive source and when I hid in it I imagined myself absorbing neutrons from the glowing walls, mutating into someone else, a super heroine-astronomer – or at least someone who felt like she might belong. But I wasn't Dr Wonder Woman. Like a small comet that once started its journey in the solar system, and was now travelling outwards and away from all the source of heat and light it had ever known, I eventually left.

I joined the civil service to work in outer space policy and the regulation of the space industry. Now I was responsible for making sure that British companies didn't launch dangerous satellites that fell down onto people's heads, and that they had bought enough insurance to cover the subsequent claims if any satellites did crash and burn. The British Government's approach to the commercial satellite industry was that it was a Good Thing and should be encouraged, the regulatory process was simply to ensure safety.

I did this job in an office next to Victoria Station in central London, where I could detect each train passing underneath the building from the way that the walls vibrated. One day there was a partial solar eclipse visible from southern England, and all of us civil servants trooped outside to stand around on the street, where we squinted upwards for a glimpse of the sky between office buildings. It was cloudy, of course. But through the clouds I was still able to see the Moon moving in front of the Sun. Because it wasn't a total eclipse the sky wasn't dark enough to make the pigeons think about night-time roosting. Nevertheless the sight of the Sun as a temporary crescent was genuinely uncanny, it did not go with Government office blocks or tourists trying to find Westminster Abbey at the wrong end of Victoria Street, or the sweet-rancid smell of old doughnuts emanating from the nearby coach station. The eclipse affected the quality of the daylight, flattening it out in a way that reminded me of the flimsiness of my surroundings, as if everything was just a stage set and the real events were taking place far away.

During the time that I spent doing this job, a year that was book-ended by autumnal equinoxes, I read about 'auctions' of parts of the electromagnetic spectrum to satellite companies so they could transmit and receive telecommunications at specific frequencies without interference from other companies. I read about the United Nations Office for Outer Space Affairs keeping a log of all the satellites launched around the world, a log that contained their positions and orbits above the Earth. I read glossy brochures published by companies about their plans to launch future satellites that would revolutionise the way we talked to each other. But I found it hard to imagine this new world of 'upstream' suppliers and 'downstream' applications, getting confused by this terminology and too often failing to understand that it was a metaphor, and persistently imagining a river flowing and bearing along with it pieces of metal, antennae, satellite dishes all being washed away in a great and incoherent water of babbling electronic voices. A technological version of the Orpheus myth; after his death and dismemberment his head floated along the river Hebrus, still singing as it went, and accompanied by his lyre which then sailed up to heaven where it can be seen as one of the constellations, Lyra.

The Government's legal power to require that the satellite companies act responsibly stemmed from the Outer Space Treaty, drafted by the United Nations around the time of the Apollo programme in the mid 1960s. This treaty states that anyone is free to use outer space for peaceful activities, but also that space belongs to nobody; nation states are explicitly prohibited from staking a claim to any object situated in it. The American flag left on the Moon by Neil Armstrong and Buzz Aldrin does not, on the face of it, constitute a claim to lunar territory.

The Outer Space Treaty is utopian in intent, aiming to stop the enclosure of the celestial commons by would-be landowners. It's inspired by the Antarctic treaty, which allows countries to build bases on the polar continent, provided they do so for scientific purposes only, and do not claim any territory for their own. This Antarctic treaty in turn derives from the law of the high seas; anyone with peaceful intent can sail a boat in international waters.

The night sky is not just a backdrop, it's necessary for us to understand our place in, and relationship to, the cosmos – both scientifically and also culturally. But this vision is being threatened. Now, when I stand outside at night and let my eyes become dark-adapted, I'm not just battling against light pollution generated nearby. The view above me is constantly interrupted by lines of light; busy little satellites are zipping to and fro across the dark sky. Artificial satellites have been looking down at us since the Soviets first launched Sputnik in 1957, but more recently they have begun to interfere with astronomy. Now, images of distant galaxies are frequently criss-crossed with light like the glittering bars of a cage. And we are all trapped inside.

Elon Musk's SpaceX company is planning to launch 65,000 communications satellites in the next few years. It is just one of several companies aiming to establish orbiting networks of small satellites in order to improve communications around the globe and provide cheap internet access, particularly in developing countries unable to install expensive earth-bound infrastructures of cables. Such satellites can be democratizing, it is important that the citizens of these countries don't get left behind in the information revolution. But there is no international body that coordinates plans to launch satellites, no effective regulatory system capable of preventing anyone with enough money and ambition from cluttering up our sky with hundreds and thousands of pieces of metal.

These groups of satellites are referred to in the space and telecoms industries as 'constellations', and I worry that soon these artificial assemblages will block our views of Andromeda or Lyra. All constellations are human constructs. The group of stars we name 'Orion' have no direct physical connection with each other, they are lightyears apart and what we identify as a definite configuration is just a line-of-sight effect, made coherent by our desire to see patterns in randomness. But these patterns give us a means of ordering what we see, of cataloguing our observations. They also help us to remember tales passed down to us from previous generations. They are a means of collecting stories and knowledge.

It is not just the working satellites that photobomb astronomical images. These machines do not last for ever; they require an internal supply of gas with which to manoeuvre and position themselves in their orbits and when this gas runs out, it cannot be replenished. A satellite typically lasts for about ten years, and after that, it has no more use and is obsolete. The ones nearest to us in low Earth orbits may break up and crash back down to the ground, while those further away simply hang around as debris. The European Space Agency (ESA) has estimated there are more than 28,000 pieces of debris orbiting the Earth that are currently being tracked by space surveillance networks, and the combined mass of all artificial objects orbiting the Earth is greater than 9000 tonnes. Just as the surface of the Earth is being laid waste, so is the space above us being turned into a junkyard.

This debris is not just caused by satellites. The vast majority of rockets that launch these satellites are 'expendable launch vehicles' that can only be used once. After their launch they break up but continue to orbit the Earth as spent casings and boosters. ESA has itself contributed to much of the debris circling the Earth, through its use of Soyuz and Ariane rockets launched from French Guiana in South America.

Elon Musk has stated that the main purpose of launching these commercial satellites is to generate a profit to pay for future human missions to Mars. Low earth orbit is being used as a stepping stone to create more distant colonies.

The frames of rockets are usually made from strong and lightweight metals, most commonly aluminium or titanium, mined in countries such as China, South Africa and Mozambique. Rockets are presented to us as the technology of future hopes, a way of launching communication tools and scientific experiments, and perhaps future human colonies, into outer space. But the origin of those rockets, the metals of which they're constituted, come from below the surface of the earth and have to be dug out – most frequently by hard labour. In 2012 a group of miners who worked in the Marikana mine in South Africa went on strike for better pay and conditions. While mining for platinum, a metal used as a catalyst in rocket engines, they were routinely exposed to dangerous conditions such as falling rocks and hazardous dust, and were poorly paid. During that strike, 34 miners were shot dead by police, the single worst use of force by the authorities against civilians in South Africa since the end of apartheid.

These miners' work is hidden underground, and largely invisible. But what they dig out of the earth has its origins in the stars. The Big Bang created the first atoms of the three lightest elements; hydrogen, helium and lithium. Heavier elements are formed in the cores of stars which act like vast nuclear fusion plants, helium atoms are squeezed together under enormous pressures until they combine to make carbon. Atoms of even heavier elements such as platinum, aluminium and titanium are created at the moment of a star's death and flung out by the subsequent explosion into the surrounding space where they may land on nearby planets.

An atom of metal has its origin in a star before being buried in the Earth's crust, to be dug out and become an essential component of a rocket that flies into space, and from there transformed into a piece of debris orbiting the Earth.

The Outer Space Treaty's prohibition on 'nation states' claiming territories does not extend to private companies, and the Treaty is silent on who may legally own the material extracted from an object in outer space. This might sound like science fiction, or as theoretical as general relativity, in fact it has already happened. In 2003 the Japanese Aerospace Agency JAXA launched the Hayabusa spacecraft that briefly landed on the nearby asteroid Itokawa, scooped up some grains of rock from its surface and then returned them to Earth. In 2014 ESA landed the Philae explorer onto the tiny comet Churyumov-Gerasimenko, and information transmitted back to Earth helped us understand the origin and make-up of this comet. The NASA spacecraft OSIRIS-REx landed on the near-Earth asteroid Bennu, collected some samples and dropped them off in September 2023, before making its way to another asteroid.

Now private companies such as the Asteroid Mining Corporation and AstroForge also want to land on asteroids in the next few years, and mine them for rare metals such as gold and platinum. There are more than 70 private missions to the Moon planned in the next decade, not just to carry out mining but also to establish data processing centres. The data cloud will no longer appear to hover in some unspecifiable place, it will be located on the Moon because this is being seen as a safer place than the Earth for our valuable information. These commercial ambitions are being encouraged by some countries, in particular USA and Luxembourg, which have both passed laws allowing private companies incorporated in their territories to own any resources they are able to extract from space. These countries are enabling the commercial exploitation of space.

Outer space is shifting ontologically. It's no longer a location of religious awe, or an entity to be understood through the medium of scientific observations. Now it is simply something to be used or exchanged for other goods. It is increasingly being viewed as capital. And it's not just the objects in space, but also the empty space itself that is becoming capital; 'low earth orbits' are occupied by satellites that enable high speed trading between different stock exchanges around the world, ensuring that the time delays after a request is sent from one financial centre to another are minimised, in order to maximise the profit.

Perhaps it doesn't do to be too sentimental about historical space exploration carried out by nation states rather than profit-seeking companies. The Apollo missions were triggered by American anxieties over the Soviets gaining the upper hand in the 'space race', which was simply an adjunct to the Cold War. The Saturn rockets used in the Apollo programme were developed by the German Wernher von Braun and can trace their technological DNA back to the Nazis' V2 rockets. There has always been a queasily intimate link between space-based telescopes and warfare; the Russian commercial rockets Soyuz, regularly used by ESA to launch its scientific experiments, are simply repurposed inter-continental ballistic missiles. Until it was put on ice following the Russian invasion of Ukraine, this business relationship between ESA and the Russian Government was frequently presented as a peace dividend from the end of the Cold War.

The digital cameras on board space-bound telescopes such as the Hubble (and also those used with ground-based telescopes), are essentially similar to military and spy cameras. The Lovell radio dish at Jodrell Bank was partially constructed from the gun turrets of British Navy destroyers which had seen action in both World Wars.

However, with Luxembourg's 'European Space Resources Innovation Centre' and its stated aims to enable mining on asteroids and comets, we are entering a new stage. These proposed activities threaten not only other astronomical bodies, but also our own home. Some nascent asteroid mining companies have argued that by mining *there* we're protecting the environment *here*, that we'll take some pressure off the Earth's resources if we can go elsewhere to mine essential materials without any apparent consequences for life on our planet. But it will have the opposite effect. It will simply encourage the view that it doesn't matter what we do or where we go, because there will always be new places to exploit. At least on our own planet, we can in principle document and regulate the damage caused by industrial mining, and we can actually do this using satellites themselves to monitor illegal and destructive land use. But this will not be the case elsewhere on asteroids or comets. On these tiny and remote objects, any company will be able to mine with impunity and without oversight.

We need to reclaim gazing at the night sky. Over the past thirty years astronomy has been transformed by technology from a data-poor to a data-rich subject. We have grown used to accessing aesthetically dazzling images that also tell us fundamentally important facts about the Universe. But perhaps we have collectively forgotten that we don't actually need any technology at all to be astronomers. Tycho Brahe's revolutionary observations of Mars were made before the telescope was invented. We can see and learn so much simply by using nothing more, or less, than our eyes.

To look at the stars is no longer passive, it is an active, political act. To stand outside on a clear night with your head tilted back is to engage with the last – and the largest –wilderness.

During my work on regulating outer space, the Government was approached by an American company who wanted to use an island in the Caribbean as a rocket launching site. This unpopulated Sombrero Island was just off the coast of Anguilla and came under its jurisdiction. So why had the Americans contacted us and not the Anguillan Government? Because Anguilla is a British Overseas Territory, formerly part of the British Empire, and although it now has its own government the UK Government still exercises control over foreign policy. Deciding whether or not rockets can be launched from Anguilla was part of that control; space exploration as a vestigial colonial activity. The Anguillan Government were all in favour of this new launch site (they would be paid for each launch by the American company), the British less so. The Foreign Office had just published a paper on encouraging environmental diversity amongst the British Overseas Territories, and Sombrero is famous for its seabirds, in particular different species of terns and gulls. Giving permission for a commercial rocket launch pad that would tarmac most of the island and destroy birds' nesting sites did not really seem consistent with the Foreign Office's new policy. But the UK Government refused to say 'no' outright, not wanting to appear opposed to a commercial space project and correctly guessing that all they had to do was stall for time until the Americans lost interest, and went elsewhere to a more amenable country.

But the colonial aspects to British space ambitions are still hanging around, refusing to go away. Now, there are plans to build similar launch sites in Britain itself, ones suitable for small rockets to carry satellites (mostly ones destined to go into polar orbit, to be used for land monitoring and weather forecasting). The proposed locations are in Cornwall, Shetland and Sutherland, the latter is likely to see the first commercial rocket flights from the UK. The future Sutherland Spaceport will be situated on the A'Mhoine peninsular on the north coast of Scotland near the Kyle of Tongue. This part of Scotland, called the 'Flow Country' because of its peat bogs, is spectacular, wild and seemingly almost empty; the nearest village to the proposed port is Talmine, about ten kilometres away. This is apparently an ideal place for a spaceport, it is far away from any built-up areas and so any accidents are very unlikely to have consequences for human life; exactly the same argument that was used to justify the building of the UK's first nuclear power station Dounreay in the 1950s, fifty kilometres on the coast due east of A'Mhoine.

I have taken the train from Inverness north to Thurso many times. As I look out of the window at this apparently pristine landscape, my eyes snag on piles of stones. These structures appear prehistoric, in fact they are much more recent. They are all that remains of the ruined crofts, deliberately destroyed by having their roofs set on fire to 'dissuade' the tenants from trying to stay. The line that you can draw from the Sutherland Clearances (perhaps the most extreme version of the Clearances that happened all over Scotland) to the development of the spaceport in a place that is conveniently under-inhabited, is straighter than this railway line. The present-day population of Sutherland is under 13,000, half of what it was in 1851 when the local landowner, the Countess of Sutherland, enclosed the common grazings, expelled the tenants and replaced them with the new Cheviot breed of sheep, which was much larger than the traditional Highland breed and therefore provided more profit, but at the cost of needing more land to graze on. Sheep provided landowners with more money than farmers, therefore the farmers had to go. It was a straightforward economic calculation.

The tenant farmers, stripped of their rights to the hitherto common land, to their animals, to their ancestral homes, were 'encouraged' to become fishermen, even though they had no skills or desire to do so. Many of them moved down south to cities where they became part of the workforce in new factories that were being established at the beginning of the Industrial revolution, or they emigrated to Canada and USA. Thus, there is a confluence between the expulsion from the land of tenant farmers and the rise of coal power, a confluence that meets in Glasgow, where the expelled people found work in shipbuilding and cotton mills.

Some of the expelled farmers did try to become fishermen, but fishing could not support the sudden influx of people seeking work along the coasts of Scotland. For a short period of time some people were employed to collect kelp, others were paid to collect guano from seabird colonies such as Bass Rock off the coast of East Lothian. Guano was a rich source of ammonia, and essential to the production of fertiliser until the German chemist Fritz Haber worked out how to synthesise it in 1909. Chemically speaking, it is first cousin to ammonium nitrate which is used as an explosive for mining, as well as a solid fuel for rockets. Sombrero Island was also mined for guano in the late 19th century.

Picture this. The people that have been pushed off the land they had farmed for generations, expelled from the grazings and the runrigs, their stone cottages destroyed, now make their way along the coast, collecting the kelp and loading it into hessian sacks. The kelp industry is growing, useful for a source of potash which is needed to make transparent glass. But, as in the future with the guano, potash will soon be synthesised and the kelp industry will collapse as quickly as a sheep carcass once the knife is stuck into it. While the people work, they glance towards the sea. The constant motion of the water, and the way it can mark the passage of days and nights, and the strands of kelp that it leaves on the beach as a sort of clock, all this is alien to them. The birds that dive into this water, emerging with fish caught in their sharp beaks, are also to be feared, for the seagulls that are rearing their chicks on nearby cottage roofs are aggressive, swooping low to peck at people's heads.

The first woman in space was the Soviet cosmonaut Valentina Tereshkova, who made a three day-long flight in June 1963 on the Vostok 6 mission, orbiting the Earth 48 times. Her call sign was *Chaika*, the Russian word for seagull, and her first broadcast after the rocket's launch was *It is I, Seagull! Everything is fine. I see the horizon; it's a sky blue with a dark strip. How beautiful the Earth is... everything is going well.*

Everything is not going so well in orbit around the Earth nowadays, what with the wreckage of old space missions vying for position amongst the commercial telecoms satellites and the secret spy satellites.

The French space agency CNES decided to situate its space centre and launchpad at Kourou in French Guiana because that country lies on the equator, and rockets launched from there benefit from the relatively fast speed of the Earth's rotation, meaning they can reach outer space with less fuel than those launched from other sites. There is also less likelihood of damage caused by rockets malfunctioning on launch, the subsequent debris will hit the sea or unpopulated areas on land, and the population density of French Guiana is one of the lowest in South America. And rockets do malfunction, the maiden voyage of the Ariane 5 rocket in 1996 went dramatically wrong, exploding forty seconds after its launch and showering the ground with debris. CNES has an ongoing agreement with ESA to launch the latter's satellites from here, and it's one of the largest and busiest launch sites in the world.

But French Guiana is not an independent country capable of making its own choices about what sort of infrastructure to build or space activities to carry out. It's not even an autonomous or overseas territory, but rather a *département* of France whose inhabitants are technically French citizens. Many French Guianese are descendants of slaves forced to work on sugar plantations. France funds French Guiana, but much of this money goes to supporting the space centre and doesn't benefit the local people. In April 2017 the space centre was occupied for three weeks by Guianese people protesting their poor quality of life, the high cost of food and lack of roads, schools and medical facilities. They drew attention to the huge contrast between the crumbling infrastructure outside the perimeter fence of CNES and the well-funded rocket launching site within it. Ten thousand people gathered outside this fence, and when a handful were invited inside to negotiate, they refused to leave and managed to delay the planned launch of an Ariane rocket for a month.

Photos of the CNES space centre at Kourou show rockets, gantries, large hangars, the flags of the twenty-five European states that contribute to ESA, the control centre, and satellite dishes pointing at the sky. Wide-field shots show an anonymous, anywhere-land, with nothing to indicate which country, or even continent, we're looking at. Just eleven kilometres away from the rockets lies Devil's Island, a penal colony used by the French until it was shut down in 1951, at which time prisoners who could not pay their own passage back home were stranded in French Guiana.

The spaceport in French Guiana was first established in

1964, after an earlier one in Algeria was now no longer available for use (by the French) because of that country's successful fight for independence.

In its reliance on distant France, a country most of its citizens have never visited, French Guiana is a rehearsal for a future space colony. Kourou is 7000 kilometres from Paris, many times further than the International Space Station from the surface of Earth. French Guiana is a territory that is simultaneously situated in the colonial slave-owning past as well as the space-hopping future, a place where formerly transported prisoners walked the same streets as rocket engineers. It's a country that's abundantly rich in natural resources, and yet the majority of whose citizens earn less than half of the annual income of people in France itself. And if French Guiana is an Earth-bound space colony then its Devil's Island is doubly so. Its most famous inmate Captain Alfred Dreyfus spent five years in solitary confinement, much of this time chained to his bed, after being falsely accused of treason (by antisemites) and subsequently convicted.

It's possible to escape from the surface of our planet, simply by using enough fuel. But we have not yet worked out how to escape our past. Outer space is apparently waiting for us to visit, but is in fact already populated with the ghosts of people who had no choice about their own space on the surface of the Earth, no rights to it, who were exploited and forced to move from one place to another.

Is there a different way of doing space exploration? A way that avoids colonial capitalism? In 1964, just as the Apollo programme was getting underway, the 'Zambian National Academy of Science, Space Research and Philosophy' was founded by Edward Makuka Nkoloso, whose ostensible aims to send people to the Moon and Mars were greeted by ridicule in the western media; how could anyone seriously train to be an astronaut by rolling down a hillside in an oil drum? But this low-tech training wasn't completely different from that developed by the other, better known, space programs. The Soviets' equipment didn't look much more sophisticated, and yet it clearly worked, and the Americans might have had more advanced technology but they were still doing many of the essential calculations by hand.

The dissonance between the Outer Space Treaty and the military associations of Apollo and Soyuz is exposed by the very existence of the Zambian Space Academy; Nkoloso stated that while he wanted to send Zambians to Mars, he didn't want to impose nation states on the indigenous Martians.

1964 was the same year that Zambia gained its independence from Britain (and two years after Algeria gained independence from France). Nkoloso took part in the fight for liberation against the colonial UK Government in what was then Northern Rhodesia, and perhaps that is why this afronaut training program resembles those designed for freedom fighters in the colonized countries of Africa: Angola, Mozambique, Southern Rhodesia (now Zimbabwe), Algeria, Kenya, and elsewhere. Perhaps that is one reason the anti-colonial ambitions of Zambian Space Academy still resonate today as the Outer Space Treaty's ideals get chipped away by private companies planning to launch missions to mine precious metals from asteroids and the Moon.

When she wasn't constructing astrological charts, my mother was a knitwear designer who ran her own business. She ran a 'cottage industry', employing other women who all worked with their knitting machines in their own homes, making the clothes that she had designed. In her workroom on the top floor of our house, cones of Shetland wool were stacked from the floor to the ceiling. It was always a small business, my mother repeatedly refused funding from other sources to expand it, saying that she wanted to make clothes and stay in control, not make money.

The two of us argued a lot, and not just about astrology. She voted for Thatcher because she thought the trade unions had too much power and unemployed people should simply try harder to get a job. Nevertheless through her work, the actual physical labour of drawing and knitting and sewing, I could witness a different approach to private enterprise; one that cared for its raw materials and finished products, as well as for the individual workers who enabled the transformation from the former to the latter.

It's hard to see how this business model translates across to outer space, hard but not impossible. Developing countries are operating satellites for their own use rather than remain dependent on communication services supplied by richer countries. Satellites themselves are becoming smaller, which means more of them can be launched by a single rocket. Re-useable rockets are a reality; Musk's SpaceX rockets are now routinely used by NASA and other space-based companies. ESA plans to launch the Clearspace-1 clean-up mission in 2026 as a pilot mission to remove debris. The UN led the way in negotiating the Outer Space Treaty in the 1960s, it can still save the commons.

My mother died a long time ago. I have pictures of her alongside images of the quasars I once studied, and which so preoccupied me that I could envisage them as I walked through the streets of Edinburgh. After her death I became a writer so I could tell tales of people in outer space.

ACKNOWLEDGEMENTS

First and foremost I'm really grateful to Aaron Kent and all his co-workers at Broken Sleep Books who support and promote so many writers.

A portion of this essay was published in a slightly different version as 'We are not special: white pages and dark matter' in the anthology *Corroding the Now*, edited by Francis Gene-Rowe, Stephen Mooney and Richard Parker, and published by Veer Books and Crater Press.

Thanks as ever to my fellow writer-pals for their brisk feedback and kind words.

LAY OUT YOUR UNREST